CLASS 20 LOCOMOTIVES

Andrew Cole

amberley

Four members of the class were sold to CFD in France, and were based at Autun. They were overhauled and painted into blue and orange before being exported. All four were returned to the United Kingdom and, although one was scrapped, the other three still survive in preservation.

Hunslet Barclay purchased seven members of the class for weed-killing trains, and six of the class were refurbished, but the seventh member was used as a source of spares, and they were eventually sold to DRS. The highlight for the class was when three locos worked to Kosovo, as part of the train for life, working for NATO for a number of months before returning to the UK. They were all sold to Harry Needle, with just one member scrapped.

The majority of the class have now been scrapped, with most being scrapped by Vic Berry, Leicester, and M. C. Metals Springburn, but over twenty of the class survive in preservation. There is also a pool of locos allocated to main line use, being used on London Underground tube stock delivery trains from Derby.

The class has certainly stood the test of time, with locos still to be seen on the main line today. I hope you enjoy looking through this collection of photographs as much as I have enjoyed putting them together.

No. 20001, 20 September 2009

No. 20001 (D8001) stands outside one of the repair sheds at the Midland Railway Centre, having been repainted into BR blue livery. This loco is nearly sixty years old, having been built in 1957.

No. 8001, 21 November 2015

No. 8001 (20001) is seen at the Midland Railway Centre having been beautifully restored to BR green, and carries its pre-TOPS number, but without the 'D' prefix.

No. 20002, 22 June 1985

No. 20002 (D8002) rests at Saltley depot, Birmingham, waiting its next turn of duty. No. 20002 would end up scrapped by M. C. Metals, Springburn in September 1990.

No. 20002, 18 March 1990

No. 20002 (D8002) is seen condemned at Doncaster Works following withdrawal. This would eventually be sent north to M. C. Metals, Springburn for scrap, which was completed in September 1990.

No. 20005, 2 March 1985

No. 20005 (D8005) stands at Bescot depot, Walsall, along with other members of the class. When delivered, the first few members of the class had oval buffers, which No. 20005 still retains. No. 20005 was scrapped by M. C. Metals, Springburn in December 1990. A large number of the class were scrapped at Springburn.

No. 20006, 4 January 1986

No. 20006 (D8006) is seen outside the shed at Tyseley depot, awaiting repairs. Tyseley didn't have a main line allocation at this time; it was mainly used as a DMU depot. No. 20006 has lost its original oval buffers, having had them replaced with standard round buffers, and it was scrapped by M. C. Metals, Springburn in June 1991.

No. 20007, 4 April 1990

No. 20007 (D8007) is seen at Derby waiting to depart with a working to Matlock. This train was top and tailed with No. 20166 at the other end, and was formed of three Network South East Mk 2 carriages. Happily this loco was preserved at the Great Central Railway, Ruddington.

No. 20008, 29 May 1988

No. 20008 (D8008) is seen stabled at Thornaby depot in BR blue livery, but with a few local additions – these being large numbers and a kingfisher depot sticker. This was another Class 20 to be scrapped by M. C. Metals, Springburn, this occurring in November 1993.

No. 20008, 22 August 1993

No. 20008 (D8008) stands condemned at Thornaby. It had been withdrawn four years earlier, and would only last another three months, being scrapped by M. C. Metals, Springburn in November 1993.

No. 20010, 13 April 1985

No. 20010 (D8010) receives a complete overhaul at Derby Works in April 1985. As can be seen, a new engine has been fitted and the loco would emerge carrying red stripe Railfreight livery. Another Class 20 to be cut up by M. C. Metals, Springburn, this being completed in January 1994.

No. 20013, 6 May 1974

No. 20013 (D8013) trundles through Derby station running as a pair along with No. 20197 on a rake of mineral wagons. The BR double arrow symbol would later be moved onto the bodyside, rather than being on the cab side. This loco would be scrapped by M. C. Metals, Springburn in September 1993.

No. 20015, 14 August 1988

No. 20015 (D8015) sits condemned at Thornaby depot, already having had its engine removed. This loco was scrapped by Vic Berry, Leicester in December 1988.

No. 20016, 25 February 1984

No. 20016 (D8016) is seen undergoing an overhaul inside Crewe Works. This would be released still carrying BR blue livery, which it carried until withdrawal. This loco is still intact today, being owned by Harry Needle and stored at MOD Long Marston.

No. 20017, 29 October 1983

No. 20017 (D8017) is seen inside the melt shop at Crewe Works. This was the place where the locos were scrapped, although it would take over a year for the loco to be finally cut – this being completed in March 1985.

No. 20020, 3 May 1986

No. 20020 (D8020) is seen stabled at Saltley depot. This was the first Class 20 built by Robert Stephenson & Hawthorn, and is seen complete with Motherwell flying salmon depot sticker. No. 20020 was later preserved at the Bo'ness and Kinneil Railway.

No. 20023, 14 March 1987

No. 20023 (D8023) is seen stabled at Bescot carrying red stripe Railfreight livery. This particular loco was renumbered to 20301 for Peak Forest workings for six months in 1986, but was renumbered back to 20023. This was scrapped by M. C. Metals, Springburn in April 1992.

No. 20026, 2 April 1983

No. 20026 rests at Bescot depot, Walsall carrying BR blue livery. This loco was scrapped by M. C. Metals, Springburn in September 1991.

No. 20029, 9 May 1985

No. 20029 (D8029) is seen shunting round Saltley depot having been refuelled. This was another Class 20 to be scrapped by M. C. Metals, Springburn, being cut in September 1993.

No. 20030, 6 June 1987
No. 20030 (D8030) is seen on display at Worksop Open Day 1987. This loco had, along with No. 20064, been returned to BR green livery, and had been named *River Rother*. This would be scrapped by M. C. Metals, Springburn in August 1991.

No. 20031, 14 February 1987
No. 20031 (D8031) is seen stabled at Bescot depot. At this time at weekends, there used to be long lines of stabled Class 20s based at Bescot. No. 20031 was preserved, and can be found at the Keighley & Worth Valley Railway.

No. 20032, 13 April 1985
No. 20032 stands at Toton depot, Nottingham, its home depot at the time. This loco spent many years in long-term storage, but was finally scrapped by European Metal Recycling, Kingsbury in February 2012.

No. 20041, 24 February 1980
No. 20041 (D8041) stands at Saltley along with other members of the class. This loco was unusual in that the running number was on the bodyside, rather than the cab side. No. 20041 was renumbered to 20904 by Hunslet Barclay for weed-killing trains, and was eventually sold to DRS.

No. 20042, 1 January 1987

No. 20042 (D8042) stands on its own at Bescot on New Year's Day 1987. Upon withdrawal this loco was sold to Waterman Railways, who painted it black, but it was eventually sold to DRS, who rebuilt it as No. 20312.

No. 20045, 13 April 1985

No. 20045 (D8045) stands on the fuel line at Toton depot carrying BR blue livery. This loco was scrapped by M. C. Metals, Springburn in October 1991.

Class 20 Locomotives

No. 20046, 29 October 1983

No. 20046 (D8046) is seen at Crewe Works having just been overhauled, and is waiting a return to its home depot of Tinsley. M. C. Metals, Springburn would scrap No. 20046 in September 1993.

No. 20047, 10 July 1994

No. 20047 is seen stabled inside the RFS works at Doncaster. This had been withdrawn from BR service in 1991, and was sold to RFS for Channel Tunnel construction workings. RFS renumbered it 2004, and it was later sold onto DRS and was rebuilt as 20301.

No. 20049, 26 December 1985

No. 20049 (D8049) spends Christmas 1985 stabled at Saltley depot, Birmingham. This would eventually be scrapped by Vic Berry, Leicester in August 1988.

No. 20050, 28 July 1984

No. 20050 (D8000) is seen condemned at Doncaster Works in July 1984. This was the first Class 20 to be built, being completed by English Electric in 1957, and had been withdrawn from BR service in 1980. It was earmarked for preservation by the National Railway Museum, and was at Doncaster for overhaul and repainting back to original condition.

No. 20053, 27 October 1984

No. 20053 (D8053) receives an overhaul at Crewe Works, along with other classmates. This was withdrawn in October 1990, and was scrapped by M. C. Metals, Springburn in September 1991.

No. 20056, 9 July 2006

No. 20056 (D8056) is seen at Barrow Hill in Corus livery, and carrying their number 81. This loco is owned by Harry Needle, and is one of two used by Tata Steel at their Scunthorpe Works, No. 20066 being the other.

No. 20057, 20 November 1994

No. 20057 (D8057) is seen stored at Springs Branch depot, Wigan. This would eventually find its way to the Midland Railway Centre, Butterley for preservation.

No. 20059, 4 August 1996

No. 20059 (D8059) is seen withdrawn at Margam depot, South Wales carrying red stripe Railfreight livery. This was renumbered 20302 for a time in 1986 for Peak Forest workings, but was soon renumbered back to 20059 and spent over six years dumped at Margam before being saved for preservation. It can be found at the Severn Valley Railway.

No. 20059, 8 February 1986

No. 20059 (D8059) rests at Saltley in red stripe Railfreight livery. Two months later it was renumbered 20302 for Peak Forest workings, but would only carry the number for seven months, being renumbered back to 20059.

No. 20063, 25 May 1986

No. 20063 (D8063) is seen stabled at Grangemouth depot, Scotland, complete with Eastfield Scottie dog sticker. This was later sold to CFD, France and was exported to Autun in France for departmental work, but was repatriated back in 2005.

No. 20063, 24 June 2000

No. 20063 (D8063) is seen inside the shed at Autun, France carrying CFD number 2002. This was sold to CFD, Chemins de Fer Departmentaux, for departmental work, and was one of four such locos sold to France. All four were later returned to the UK.

No. 20064, 31 May 1987

No. 20064 (D8064) is seen at the Coalville open day 1987, carrying a version of BR green livery that had been applied by Tinsley. It had gained the name *River Sheaf*, but this was later removed. No. 20064 was scrapped by M. C. Metals, Springburn in December 1990.

No. 20065, 13 April 1985
No. 20065 (D8065) is seen stabled at Toton in BR blue livery, along with many other classmates. This was another Class 20 to be scrapped by M. C. Metals, Springburn, this being completed in January 1991.

No. 20066, 8 February 1987
No. 20066 (D8066) stands at Bescot depot, Walsall along with other Class 20s, plus one of their Class 58 replacements. This was sold to RFS for Channel Tunnel construction workings before eventually being sold to Harry Needle. It is currently in use at Tata Steel, Scunthorpe.

No. 20070, 24 May 1987

No. 20070 (D8070) rests at its home depot of Thornaby. No. 20070 had received the name *Leyburn*, and also gained a Thornaby Kingfisher. No. 20070 was sold to M. C. Metals, Springburn for scrap, which they completed in December 1992.

No. 20071, 27 October 1984

No. 20071 (D8071) is seen inside the paint shop at Crewe Works, having been rubbed down ready for a new coat of BR blue livery. This was another Class 20 sold to M. C. Metals, Springburn for scrap, which was completed in March 1995.

No. 20072, 27 August 1995
No. 20072 (D8072) is seen on display at Basford Hall open day. This would eventually go to MOD Long Marston for long-term storage.

No. 20073, 17 May 1981
No. 20073 (D8073) is seen stabled at Saltley in BR blue livery, and has lost its headcode discs. No. 20073 would eventually be sold to Harry Needle, but it was sent to C. F. Booth, Rotherham for scrap, which was completed in April 2006.

No. 20075, 27 August 1995

No. 20075 (D8075) is seen at Worcester open day carrying British Rail Telecommunications livery. There were four Class 20s repainted in these colours, and all received names, No. 20075 being *Sir William Cooke*. No. 20075 was sold to DRS who rebuilt it as No. 20309.

No. 20077, 18 March 1990

No. 20077 (D8077) stands condemned at Doncaster Works carrying red stripe Railfreight livery. It had received the damage at Bickershaw Colliery and was sent to Vic Berry, Leicester for scrap the following month.

Class 20 Locomotives

No. 20078, 4 July 1987
No. 20078 (D8078) stands at Bescot at the head of a line of other Class 20s. This loco spent many years based in Scotland and would return there for scrapping, being cut up by M. C. Metals, Springburn in October 1993.

No. 20081, 20 February 1985
No. 20081 (D8081) passes through Nuneaton with No. 20113 at the head of a rake of coke wagons. No. 20081 currently languishes at MOD Long Marston, owned by Harry Needle.

No. 20081, 15 June 1985

No. 20081 (D8081) is seen passing Saltley light engine with another Class 20. No. 20081 was eventually sold to Harry Needle and is currently in long-term storage at MOD Long Marston.

No. 20081, 18 March 1990

No. 20081 (D8081) stands on the buffer stops at Toton along with No. 56007 and No. 58050. It is hard to believe all three of these locos are out of service, with No. 20081 stored at MOD Long Marston, No. 56007 stored at Leicester and No. 58050 stored at Albacete, Spain.

No. 20084, 10 April 1987

No. 20084 (D8084) is seen stabled at Saltley in BR blue livery. This Class 20 was fitted with extra fuel tanks at Crewe Works, which can be seen just in front of the cab. No. 20084 was sold to DRS and was rebuilt as No. 20302.

No. 20085 and No. 20059, 18 March 1990

No. 20085 (D8085) and No. 20059 (D8059) are seen passing Toton with an MGR working. No. 20085 was sold to RFS as a source of spare parts for their Channel Tunnel fleet, and was eventually sold to a private contractor who broke it up at Castle Donington in September 1994.

No. 20086, 8 February 1987

No. 20086 (D8086) is seen stabled at Bescot along with other members of the class. No. 20086 was sold to M. C. Metals, Springburn for scrap, which they completed in February 1991.

No. 20087, 17 May 1980

No. 20087 (D8087) is seen undergoing repairs at Toton depot, Nottingham. It had yet to have the BR double arrow moved to the bodyside. No. 20087 was sold for preservation at the East Lancashire Railway.

No. 20087, 15 July 2007

No. 20087 (D8087) is seen at Barrow Hill having been preserved at the East Lancashire Railway. The addition of snowploughs really sets the appearance off nicely.

No. 20088, 4 February 1989

No. 20088 (D8088) rests at Saltley in unbranded Railfreight livery. This was the only Class 20 to receive this livery in service, although some members have received the livery in preservation. No. 20088 was sold to RFS for their Channel Tunnel trains, and was renumbered 2017. It is currently owned by Harry Needle and is in long-term storage at MOD Long Marston.

No. 20089, 20 March 1988
No. 20089 (D8089) stands condemned at Immingham. It would sit at Immingham for another five years before being sent to M. C. Metals, Springburn for scrap, which was completed in December 1993.

No. 20090, 16 September 1985
No. 20090 (D8090) sits at Saltley having not long been released to traffic carrying red stripe Railfreight livery. This livery certainly brightened up the appearance of these locos. No. 20090 was sold to M. C. Metals, Springburn for scrap, which was completed in March 1995.

Class 20 Locomotives

No. 20092, 15 July 2007

No. 20092 (D8092) is seen at Barrow Hill during a special Class 20 event. No. 20092 carries Technical Services livery, and was one of only three locos to carry this livery – No. 20169 and No. 47972 being the others.

No. 20093 and No. 20054, 28 July 1984

No. 20093 (D8093) and No. 20054 (D8054) are seen arriving at Doncaster station with a railtour in connection with the works open day. Both locos would find their way to M. C. Metals, Springburn for scrapping. No. 20093 was scrapped in May 1992, while No. 20054 was scrapped in October 1991.

No. 20093, 20 March 1988

No. 20093 (D8093) stands outside Frodingham depot along with a Class 31 and a Class 56. No. 20093 would eventually be sold to M. C. Metals, Springburn who scrapped it in May 1992.

No. 20094, 25 February 1984

No. 20094 (D8094) is seen at Crewe Works having completed an overhaul. This was an unlucky loco, having been sold to the Great Central Railway, Ruddington for preservation, only to be sold to DRS for spares. The remains were scrapped in May 2004.

No. 20095 and No. 20029, 14 June 1990

No. 20095 (D8095) and No. 20029 (D8029) are seen passing Washwood Heath with a departmental working. Both carry BR blue livery, and No. 20095 was sold to RFS for Channel Tunnel construction workings, gaining their number 2020. It was later sold on to DRS, and was rebuilt as No. 20305 and is still in use today. No. 20029 was not so fortunate, being scrapped by M. C. Metals, Springburn in September 1993.

No. 20096, 16 October 2005

No. 20096 (D8096) is seen at Barrow Hill while on passenger shuttle duty. No. 20096 was saved for preservation at the SYRPS Meadowhall, and when that site closed it moved to Barrow Hill. It is seen carrying Railfreight livery, a scheme it didn't carry in traffic. It has now been main line certified, and can be seen mainly on London Underground delivery trains.

No. 20096 and No. 20905, 9 May 2006

No. 20096 (D8096) and No. 20905 (D8325) are seen coupled together outside Doncaster Works. Both carry Railfreight livery, a scheme neither carried in BR service. Both are owned by Harry Needle and were on hire to Advenza. Both are still in service today, but carry different liveries; No. 20096 carries BR blue, while No. 20905 carries GBRf livery.

No. 20096 and No. 20314, 31 December 2012

No. 20096 (D8096) and No. 20314 (D8117) are seen at Derby station. They were in use on London Underground delivery workings, with the bogie tank wagon coupled to No. 20314 being a barrier wagon. No. 20096 carries BR blue livery, while No. 20314 carries HNRC orange livery.

Class 20 Locomotives

No. 20098, 8 February 1987

No. 20098 (D8098) rests for the weekend at Bescot carrying BR blue livery. It had lost most of the headcode discs off the nose, retaining only the top disc. This loco would have a happy future, being preserved at the Great Central Railway, Loughborough.

No. 20099, 27 August 1985

No. 20099 (D8099) stands at the buffer stops at Nuneaton station along with No. 86001. This siding has been removed now that the platforms have been lengthened. No. 20099 was sold to M. C. Metals, Springburn for scrap, which they completed in August 1993.

No. 20102 and No. 20092, 11 April 1990

No. 20102 (D8102) and No. 20092 (D8092) are seen arriving at Derby station with a single observation coach carrying Inter City livery. Both 20s carry BR blue livery and both are still intact today, with No. 20102 being sold to DRS for rebuilding as No. 20311, while No. 20092 is in long-term storage at MOD Long Marston.

No. 20103, 28 July 1984

No. 20103 (D8103) stands at a sunny Tinsley on the day of the Doncaster Works open day. No. 20103 carries a fresh coat of BR blue livery, which had been applied at Glasgow Works, the telltale sign being the larger numbers. At least another six Class 20s can be seen in the background. No. 20103 was sent to M. C. Metals for scrap, which they completed in June 1992.

No. 20103, 18 March 1990

No. 20103 (D8103) is seen stabled at Toton carrying BR blue livery, and still retains a set of snowploughs. No. 20103 only had another year in service before it was withdrawn and sold to M. C. Metals, Springburn for scrap, which happened in June 1992.

No. 20104, 1 March 1986

No. 20104 (D8104) stands at Tyseley carrying red stripe Railfreight livery, along with similarly liveried No. 20023. Tyseley didn't have a main line allocation at the time, but locos used to visit for repairs. No. 20104 was eventually sold to DRS and was rebuilt as No. 20315.

No. 20104, 8 March 1986

No. 20104 (D8104) is seen stabled at Bescot carrying red stripe Railfreight livery. This livery certainly brightened up the appearance of the class compared to the BR blue livery. No. 20104 was eventually sold to DRS, who rebuilt it as No. 20315.

No. 20105, 13 May 1988

No. 20105 (D8105) runs nose first past Washwood Heath with an engineer's working, complete with guard's brake van. It was unusual to see the 20s running singularly, nose first. No. 20105 was sold to RFS for Channel Tunnel construction workings, and received their number 2016. No. 20105 would eventually be stripped for spare parts at Barrow Hill and was scrapped by T. J. Thomson, Stockton.

No. 20105, 15 November 2008

No. 20105 (D8105) stands condemned at Barrow Hill. No. 20105 carries RFS livery and their number 2016. It would eventually be stripped for parts and was sold to T. J. Thomson, Stockton for scrap.

No. 20108, 14 March 1987

No. 20108 (D8108) rests at Bescot in red stripe Railfreight livery, complete with snowploughs. This loco spent most of its life north of the border, only coming south for the last three years of service. No. 20108 was sold to RFS for Channel Tunnel construction workings and was renumbered No. 2001. It was later sold to Harry Needle for spare parts, and was scrapped by European Metal Recycling, Kingsbury in October 2005.

No. 20108 and No. 20215, 3 May 1990

No. 20108 (D8108) and No. 20215 (D8315) are seen running through Derby station, both carrying red stripe Railfreight livery. Of note is the guard's brake van, which is also included in the MGR rake. Both locos have since been scrapped, No. 20108 was scrapped by European Metal Recycling, Kingsbury in October 2005, whereas No. 20215 was cut by C. F. Booth, Rotherham in November 2009.

No. 20110, 23 May 2009

No. 20110 (D8110) is seen on display at the Eastleigh Works open day, 2009. No. 20110 was another Class 20 to spend most of its life in Scotland, only coming south to Tinsley in 1986. No. 20110 was preserved by the South Devon Railway at Buckfastleigh and is seen having been beautifully restored to BR blue livery, complete with snowploughs and tablet catching recess on the cabside.

No. 20113, 20 February 1985

No. 20113 (D8113) is seen paired with No. 20081 at Nuneaton with a rake of coke wagons. No. 20113 was another Class 20 that was sold to RFS for Channel Tunnel construction workings, this time being renumbered 2003. When these workings were completed, it was sold to DRS as a source of spares for their Class 20/3 fleet and was then sold to Harry Needle for final scrapping, which took place opposite the DRS depot at Carlisle in December 2003.

No. 20114, 18 March 1990

No. 20114 (D8114) is seen withdrawn at Toton, complete with snowploughs, being stripped of spare parts to keep other Class 20s in service. No. 20114 was another to spend almost all its working life in Scotland, only coming south in 1989 to spend nine months at Toton before final withdrawal. It was sent back to Scotland to be scrapped, M. C. Metals, Springburn cutting the loco in July 1991.

No. 20116, 13 August 1985

No. 20116 (D8116) is seen stabled at Saltley depot, Birmingham carrying BR blue livery and also retaining an Eastfield Scottie dog sticker, despite being allocated to Tinsley at the time. This would be withdrawn with collision damage and was sold to Vic Berry, Leicester for scrap, which was completed in December 1988.

No. 20117, 27 August 1995

No. 20117 (D8117) is seen on display at Crewe Basford Hall open day, despite being withdrawn. This loco was sold to DRS and was included in their Class 20/3 rebuild programme, becoming No. 20314.

No. 20118, 29 May 1988

No. 20118 (D8118) is seen stabled at Thornaby in red stripe Railfreight livery. No. 20118 had been named *Saltburn-by-the-Sea* by this point and also carries a Kingfisher depot sticker, underneath which is a special plaque stating that the loco has been adopted by the Cleveland Young Ornithological Group. No. 20118 has been restored to main line condition and can be seen hauling London Underground tube stock delivery trains.

No. 20118, 12 July 1992

No. 20118 (D8118) stands at Tinsley carrying red stripe Railfreight livery, although the bodyside doors have been replaced with blue doors. This was one of three Class 20s present on the day in this condition, No. 20137 and No. 20165 being the other two. No. 20118 was sold to Harry Needle and has been beautifully restored with red stripe Railfreight livery, and passed for main line use.

No. 20118, 17 November 2012

No. 20118 (D8118) is seen at Barrow Hill having been restored and repainted into red stripe Railfreight livery. It has also had its name, *Saltburn-by-the-Sea*, reaffixed and is currently passed for main line use and used on the London Underground tube stock delivery workings from Derby. No. 20118 is owned by Harry Needle.

No. 20119, 15 July 2007

No. 20119 (D8119) is seen stored at Barrow Hill. Despite being at Barrow Hill, this loco would be stripped for spare parts before being sent to C. F. Booth, Rotherham for scrap, which was completed in February 2009. Note how the loco has the tablet-catching recess in the cabside.

Class 20 Locomotives

No. 20121, 16 October 2005

No. 20121 (D8121) stands in storage at Barrow Hill wearing BR blue livery, and is complete with snowploughs. This was another Class 20 to spend many years working in Scotland, only to come south in 1984. Despite the appearance, the loco has since been undergoing a restoration and is currently at Barrow Hill carrying the latest Harry Needle orange livery.

No. 20122, 29 May 1988

No. 20122 (D8122) is seen at Thornaby carrying red stripe Railfreight livery. No. 20122 has had two Kingfisher depot stickers applied, a white one and a painted one. Also note that the Railfreight logo on the cabside is cast rather than sticker. No. 20122 carries the name *Cleveland Potash*, which it only carried for about three years. The loco was withdrawn in 1991 and sold to M. C. Metals, Springburn for scrap, and was cut in October 1993.

No. 20123, 20 July 1986

No. 20123 (D8123) is seen stabled at Tyseley depot, Birmingham, awaiting repairs. The loco carries BR blue livery and also has a Motherwell salmon depot sticker on the bodyside. This loco spent most of its life in Scotland, only coming south for the last eleven months of service, and was withdrawn in 1987. It was sold to Vic Berry, Leicester for scrap, which was completed in August 1988.

No. 20124, 2 August 1986

No. 20124 (D8124) stands at Tyseley waiting repairs just a month after being transferred down from Scotland to Toton and still carries a ML depot sticker on the cabside. No. 20124 would be withdrawn in 1991 and was sent to M. C. Metals, Springburn for scrap. It was cut in October 1993.

Class 20 Locomotives

No. 20127, 4 April 1990

No. 20127 (D8127) passes round the back of Derby station, nose first, with an engineer's working. No. 20127 was the last class member to be built with headcode discs, as No. 20128 was renumbered from D8050, and from No. 20129 onwards locos were built with headcode boxes. No. 20127 was another Class 20 to be sold to RFS for Channel Tunnel workings and was renumbered to 2018. It would then be sold to DRS for inclusion in the Class 20/3 rebuild programme, becoming No. 20303 in 1996.

No. 20128, 15 November 1986

No. 20128 (D8050) stands in the autumn sunshine at Bescot. No. 20128 was renumbered out of sequence, as its pre-TOPS number was D8050, and became the last numerically to be fitted with headcode discs. This loco would go on to be taken over by the British Rail Telecommunications division and would receive the name *Guglielmo Marconi*. No. 20128 would be sold to DRS and would be rebuilt as No. 20307.

No. 20129, 17 August 1985

No. 20129 (D8129) rests at Saltley carrying BR blue livery. This was the second Class 20 to be built with headcode boxes back in 1966. No. 20129 lead an uneventful life, finally being withdrawn in 1990 and being sold to M. C. Metals, Springburn for scrap, which was completed in September 1991.

No. 20131, 27 October 1984

No. 20131 (D8131) is seen undergoing overhaul inside Crewe Works. Crewe overhauled quite a few Class 20s at this time, with most having been withdrawn in 1983. No. 20131 would go on to work for British Railways Telecommunications, and would gain the name *Almon B. Strowger* before being sold to DRS for inclusion in the Class 20/3 rebuild programme, becoming No. 20306.

No. 20131, 22 May 1994

No. 20131 (D8131) is seen carrying British Rail Telecommunications livery at Worcester open day 1994. This was one of four Class 20s to receive this livery, and all four received names, this one being *Almon B. Strowger*. No. 20131 was later sold to DRS, who rebuilt it as No. 20306.

No. 20132, 13 April 1985

No. 20132 (D8132) stands outside the test house at Derby Works having just been overhauled and repainted into red stripe Railfreight livery. When clean, this livery sat nicely on these locos. No. 20132 was hired out to CTTG for use on Channel Tunnel construction workings, being renumbered CTTG 36, and was finally withdrawn from British Rail service in 1995. It was later sold to Harry Needle, who restored it to main line condition, and can be seen on London Underground tube stock delivery workings carrying the name *Barrow Hill Depot*.

No. 20133, 19 January 1985

No. 20133 (D8133) stands in the snow at Bescot. No. 20133 would later be sold to RFS for Channel Tunnel construction workings and was renumbered to 2005. Upon its return from France, it went into long-term storage until it was sold to C. F. Booth, Rotherham for scrap, which was completed in October 2002.

No. 20134, 18 March 1990

No. 20134 (D8134) is seen condemned at Toton, showing the collision damage that caused its withdrawal. It received the damage at Worksop and would be sold to M. C. Metals, Springburn for scrap, which was completed in August 1991. No. 20134 was renumbered to 20303 for just over six months in 1986 for Peak Forest workings, before being renumbered back to 20134.

No. 20137, 12 July 1992

No. 20137 (D8137) is seen at Tinsley in red stripe Railfreight livery, but with blue engine room doors. This loco once carried the name *Murray B. Hofmeyr*, but it had been removed by the time the photograph was taken. No. 20137 would be saved for preservation at the Gloucester & Warwickshire Railway.

No. 20139, 30 June 2001

No. 20139 (D8139) is seen working for CFD in France. Four Class 20s were sold to CFD for operations in France and were based at Autun, where No. 20139 can be seen. No. 20139 was renumbered to 2003, and eventually all four were returned to the UK. No. 20139 is seen standing on sleepers while its bogies are away for repairs. This is the only one of the four to have been scrapped upon return to the UK, this being completed by European Metal Recycling, Kingsbury in 2010.

No. 20140, 18 June 1980

No. 20140 (D8140) is seen stabled at Saltley depot, Birmingham carrying BR blue livery. No. 20140 lead an uneventful life, finally being withdrawn in 1993 before being sold to M. C. Metals, Springburn for scrap, which they completed in May 1994.

No. 20141, 13 July 1986

No. 20141 (D8141) is seen stabled at Bescot having just been released from works overhaul, including a repaint into red stripe Railfreight livery. No. 20141 spent all of its life allocated to the Midland Region, and at Toton, before being withdrawn in 1992. It was sold for scrap to M. C. Metals, Springburn, who completed the task in November 1993.

No. 20149, 14 February 1987

No. 20149 (D8149) is seen stabled at Bescot in BR blue livery, complete with miniature snowploughs. No. 20149 would only last another three months in service, being withdrawn in May 1987, and it was sold to Vic Berry Leicester for scrap, which was completed in October 1988.

No. 20151, 13 June 1987

No. 20151 (D8151) stands at Bescot in BR blue livery, complete with miniature snowploughs. This loco spent nearly two years in storage in the early 1980s before being reinstated to Toton. No. 20151 was finally withdrawn in 1993 and was sent to M. C. Metals, Springburn for scrap, which was completed in March 1995.

No. 20154, 4 July 1987

No. 20154 (D8154) is seen stabled at Bescot in BR blue livery, along with other members of the class. No. 20154 would be saved for preservation, first at the Churnet Valley Railway and then the Great Central Railway, Ruddington.

No. 20156, 29 May 1988

No. 20156 (D8156) is seen carrying red stripe Railfreight livery at Thornaby depot. It had received the unofficial name *HMS Endeavour* while allocated to Thornaby. No. 20156 was sold to M. C. Metals, Springburn for scrap, which was completed in October 1993.

No. 20156, 1 May 1993

No. 20156 (D8156) stands condemned at BRML Glasgow Works. It carries red stripe Railfreight livery, and has lost its unofficial name, *HMS Endeavour*. It was moved next door to M. C. Metals yard for scrap, which was completed in October 1993.

No. 20157, 12 January 1985

No. 20157 (D8157) is seen at Bescot having just been overhauled and repainted. This loco would spend its entire career allocated to Toton, from 1966 until withdrawal in 1990. It was sent to M. C. Metals, Springburn for scrap, which was completed in June 1992.

No. 20160, 30 August 1992

No. 20160 (D8160) is seen on display at Bescot open day 1992. It had been specially repainted for the open day, despite having been withdrawn eighteen months previously. It had also been named *Bescot Bo-Bo* for the event. No. 20160 was sent for scrap to M. C. Metals, Springburn, and was cut in May 1994; it must surely have been one of the cleanest locos to be scrapped.

No. 20162, 14 September 1983

No. 20162 (D8162) is seen stabled at a busy Saltley depot, Birmingham. This was another Class 20 to lead a quiet life before it was withdrawn in 1987. It was sent to Vic Berry, Leicester for scrap and was cut in December 1987.

No. 20166, 4 April 1990

No. 20166 (D8166) is seen waiting to depart from Derby station with a top and tail working to Matlock, with No. 20007 on the other end. The train consists of three Network South East Mk 2 carriages. No. 20166 was sold to RFS for Channel Tunnel construction workings and was renumbered to 2015. When it returned to the UK, it was sold into preservation at the Bodmin & Wenford Railway, but has since moved to the Wensleydale Railway.

No. 20166, 8 August 1995

No. 20166 (D8166) is seen preserved at the Bodmin & Wenford Railway. No. 20166 spent all of its life allocated to the Midland Region before being sold to RFS for Channel Tunnel construction workings in 1991. It was renumbered to 2015 and was preserved at Bodmin when it returned to the UK. While at Bodmin it received the name *River Fowey* and has since moved to the Wensleydale Railway.

No. 20168, 4 August 1996

No. 20168 (D8168) is seen stored at Margam in faded BR blue livery. No. 20168 was renumbered to 20304 for approximately twelve months in 1986 for Peak Forest workings, but was renumbered back to 20168. This loco was eventually sold to Harry Needle for further use, and has found employment at LaFarge Cement, Hope, where it carries plain white livery with a purple solebar, and the number 2. It has also been named *Sir George Earle*.

No. 20169 and No. 20147, 9 June 1983

No. 20169 (D8169) and No. 20147 (D8147) pass Saltley depot, Birmingham with a mixed freight. The first two wagons are old GWR Siphon vans. No. 20169 would go on to carry Technical Services livery and would be saved for preservation at the Wensleydale Railway.

No. 20170, 21 July 1987

No. 20170 (D8170) is seen at Saltley depot, Birmingham, having just been overhauled and repainted into red stripe Railfreight livery. The overhaul included the removal of the headcode box at the nose end and replacing it with marker lights. No. 20163 was also similarly treated. This was another Class 20 to spend all its life allocated to Toton before being withdrawn in 1991. It was sold to M. C. Metals, Springburn for scrap, which was completed in September 1993.

No. 20172, 14 August 1988

No. 20172 (D8172) is seen stabled at Thornaby depot. The loco had been spruced up by depot staff, including a red solebar, large numbers and the unofficial name *Redmire*. No. 20172 would be sold to M. C. Metals, Springburn for scrap, and it was cut in March 1995. No. 20172 was temporarily renumbered to 20305 for a while in 1986 for Peak Forest workings, but was renumbered back to 20172.

No. 20173, 24 May 1987

No. 20173 (D8173) is seen stabled at Thornaby depot carrying BR blue livery, but with red solebar and unofficial name *Wensleydale*. Thornaby always liked to personalize its locos, and the Class 20s were no exception. This loco, like many others, was sold to RFS, but only for spare parts for the rest of the Channel Tunnel fleet. No. 20173 was broken up for scrap by M. C. Metals, Springburn in February 1992.

No. 20175, 21 July 1987

No. 20175 (D8175) is seen stabling at Saltley depot, Birmingham following refuelling. It is coupled with No. 20170, and it was always nice to see a pair carrying red stripe Railfreight livery. No. 20175 carries unusually small running numbers, and when it was withdrawn from British Rail service it was sold to RFS for Channel Tunnel construction workings and was renumbered to 2007. When it returned to the UK, it was sold to DRS, but only for spares, and was subsequently sold to Harry Needle for scrap, which was completed in December 2003 opposite Carlisle Kingmoor.

No. 20176, 25 August 1988

No. 20176 (D8176) is seen passing round the back of Peterborough station coupled with another Class 20 on an engineer's train of new sleepers. No. 20176 would be sold to M. C. Metals, Springburn for scrap, which was completed in December 1993.

No. 20177, 11 April 1987

No. 20177 (D8177) rests at Bescot carrying BR blue livery, complete with miniature snowploughs. This Class 20 spent all its working life at Toton, except for three months spent allocated to Thornaby. It was stored in 1992 at Toton and spent many years dumped inside the training compound until sold to the Somerset & Dorset Loco Co. Ltd in 2001, and was used as a source of spare parts. It later moved to Tyseley, again for spare parts removal, before finding its way to the Severn Valley Railway.

No. 20178, 26 December 1985

No. 20178 (D8178) spends Christmas 1985 stabled at Saltley depot, Birmingham. No. 20178 would be withdrawn in 1989 and was sold to M. C. Metals, Springburn for scrap, being cut in June 1992. M. C. Metals certainly scrapped a large number of locos in the early 1990s.

No. 20181, 12 January 1985

No. 20181 (D8181) is seen at Saltley depot, Birmingham, carrying BR blue livery. No. 20181 was withdrawn in 1987 and moved to Doncaster Works for spares, removal and scrapping; however the shell was subsequently sold to M. C. Metals, Springburn for final scrapping, which was completed in May 1994.

No. 20182, 7 March 1987

No. 20182 (D8182) stands in a snow shower at Tyseley depot, Birmingham. This loco spent all its life allocated to the Midland Region before being withdrawn in 1991. It was sold to M. C. Metals, Springburn for scrap and it was cut in September 1993.

No. 20183 and No. 20060, 24 February 1987

No. 20183 (D8183) and No. 20060 (D8060) pass behind Peterborough station with a rake of two-axle presflo wagons. No. 20183 would eventually be sent to Vic Berry, Leicester for scrap, which was completed in March 1990, whereas No. 20060 was sold to Hunslet Barclay for weed-killing trains and was renumbered to 20902.

No. 20186, 11 June 1989

No. 20186 (D8186) is seen on display at Coalville open day 1989. This was another Class 20 to spend all its life allocated to Toton and was finally withdrawn in 1993. It was then sold to M. C. Metals, Springburn for scrap and was cut in March 1995.

No. 20187, 3 June 1993

No. 20187 (D8187) departs Doncaster for Bescot having just been released from the nearby works following overhaul and repainting into BR blue livery. This loco, along with three others, had been sold to British Rail Telecommunications, hence the BRT on the bodyside. It was later sold to DRS and was included in the class 20/3 rebuild programme, emerging as No. 20308.

No. 20188, 7 February 1988

No. 20188 (D8188) stands at Tyseley depot, Birmingham waiting repairs. This loco has lead an interesting career since it was withdrawn, being sent to Ilford training school for training purposes. It was then sold to Waterman Railways and painted in their black livery. While owned by Waterman Railways, it appeared in the James Bond film *GoldenEye* and was then sold to the Yeovil Country Railway. It was finally sold to the Severn Valley Railway.

No. 20189, 1 May 1993

No. 20189 (D8189) is seen inside M. C. Metals yard, Springburn acting as yard pilot. The loco carries M. C. Metals' livery of plain grey, and carries MC in the headcode box. It was later sold to RT Railtours and was eventually based at the Midland Railway Centre. It is currently main line registered and has carried London Transport and Balfour Beatty liveries before being painted dark blue.

No. 20190 and No. 20143, 9 November 1985

No. 20190 (D8190) and No. 20143 (D8143) are seen at Bescot, with No. 20143 having derailed. The recovery crane can be seen in the background, ready to re-rail the loco. No. 20190 would eventually be sold to DRS for inclusion in the Class 20/3 rebuild programme, emerging as No. 20310. No. 20143 must rank as one of the unluckiest Class 20 locos, being sold for preservation to the Llangollen Railway, only to be sent to M. C. Metals and scrapped in error.

No. 20190, 27 August 1995

No. 20190 (D8190) is seen on display at the Crewe Basford Hall open day 1995. Despite being on display, it was in fact in long-term storage at Crewe Electric depot, but was moved for the open day. It was eventually sold to DRS and rebuilt as No. 20310.

No. 20195, 30 December 1979

No. 20195 (D8195) rests at Saltley carrying BR blue livery. This shot shows that the BR double arrow has yet to be moved to the bodyside from the cabside, and No. 20195 still carries headcode roller blinds, rather than the domino dots. No. 20195 was sold to M. C. Metals, Springburn for scrap and was cut in March 1995.

No. 20195, 18 March 1990

No. 20195 (D8195) is seen stabled at Toton carrying BR blue livery. Toton was one of the main depots that the Class 20s were allocated to, along with Tinsley and Eastfield. No. 20195 was scrapped by M. C. Metals, Springburn in March 1995.

No. 20197, 6 May 1974

No. 20197 (D8197) is seen passing through Derby station along with No. 20013. No. 20197 was saved for preservation at the Bodmin & Wenford Railway, but the restoration was not forthcoming and the loco was sent for long-term storage at MOD Long Marston before finally being sent to European Metal Recycling, Kingsbury for scrap, which was completed in October 2011, having been withdrawn twenty years earlier.

No. 20197, April 1987

No. 20197 (D8197) is seen passing Saltley with a sister loco, having come to the aid of a failed HST. No. 20197 would be saved for preservation at the Bodmin & Wenford Railway, but would ultimately be scrapped by European Metal Recycling, Kingsbury in October 2011.

No. 20199, 21 April 1984

No. 20199 (D8199) is seen stabled at York carrying BR blue livery. No. 20199 had not long been transferred to Toton from Eastfield, and may well have been on its way south. No. 20199 would eventually be scrapped by M. C. Metals, Springburn in October 1993.

No. 20205, 20 September 2009

No. 20205 (D8305) is seen at the Midland Railway Centre, Butterley having been preserved. It carries the number 20907, although this number was never used by British Rail. No. 20209 was allocated the number for Hunslet Barclay weed-killing trains, but the renumbering never took place.

No. 20207, 25 May 1986

No. 20207 (D8307) stands forlornly inside St Rollox Works, Glasgow. It had been withdrawn following collision damage in 1983 and would only linger in the yard at Glasgow for another month, being scrapped where it stood in June 1986. Of note are the remains of another Class 20 in front of No. 20207.

No. 20211, 1 May 1993

No. 20211 (D8311) is seen condemned at BRML Glasgow Works. No. 20211 would only last another five months, being moved next door into M. C. Metals yard for final scrapping in October 1993.

No. 20212, 12 January 1985

No. 20212 (D8312) is seen stabled at Bescot in BR blue livery. At this time there were always plenty of Class 20s to be found at Bescot. No. 20212 was withdrawn in in 1991 and was sent to M. C. Metals, Springburn for scrap, this being completed in October 1993.

No. 20213, 2 June 1984

No. 20213 (D8313) is seen stabled at Bescot wearing BR blue livery. No. 20213 would be another Class 20 that was sent to M. C. Metals, Springburn for scrap and was cut in October 1992.

No. 20215, 14 June 1986

No. 20215 (D8315) is seen in red stripe Railfreight livery undergoing repairs at Tyseley depot, Birmingham. When No. 20215 was withdrawn, it was sold to DRS for use as a source of spares and was eventually broken up for scrap by C. F. Booth, Rotherham in December 2009.

No. 20217, 19 March 1987

No. 20217 (D8317) passes Water Orton running nose first with a short steel working, complete with guard's brake van. No. 20217 was withdrawn in 1989 and was sold to M. C. Metals, Springburn for scrap, which was completed in February 1991.

Class 20 Locomotives

75

No. 20223, 25 May 1986

No. 20223 (D8323) is seen stabled at Millerhill complete with Haymarket Castle depot sticker. This was another Class 20 to spend many years working in Scotland, and would also be scrapped in Scotland by M. C. Metals, Springburn in December 1993.

No. 20226, 27 October 1984

No. 20226 (D8326) is seen stabled at Cockshute Sidings, Stoke-on-Trent. When No. 20226 was withdrawn, it was sold to M. C. Metals, Springburn for scrap and was cut in June 1992.

No. 20227, 6 August 2000

No. 20227 (D8327) is seen at Old Oak Common open day carrying Metropolitan Railway maroon livery and the name *Sir John Betjeman*. This was the last Class 20 to be built, and has been saved for preservation at the Midland Railway Centre, Butterley.

No. 20227, 20 September 2009

No. 20227 (D8327) is seen preserved at the Midland Railway Centre, Butterley. It had been repainted into red stripe Railfreight livery and fitted with headlights. No. 20227 has since been passed for main line operation and can be seen on London Underground tube stock delivery trains from Derby.

Class 20 Locomotives

77

No. 20227, 21 November 2015

No. 20227 (D8327) is seen inside the repair shed at the Midland Railway Centre, Butterley. No. 20227 carries London Underground livery and is the only main line loco to do so. It is painted in this livery for use on the London Underground tube stock delivery trains.

No. 20228, 29 May 1988

No. 20228 (D8128) is seen stabled at Thornaby depot carrying BR blue livery. This was the first Class 20 built with headcode boxes, but was renumbered out of sequence, its pre-TOPS number being D8128. Of note is the fact that the headcode box has been replaced with two marker lights. No. 20228 was sold to CFD in France, along with three other Class 20s, for use on departmental trains. It was returned to the UK and has since been preserved at the Barry Island Railway, South Wales.

No. 20228, 15 July 2007

No. 20228 (D8128) is seen at Barrow Hill carrying CFD livery and their number 2004. No. 20228 had been sold to CFD, France upon withdrawal, but is seen here having returned to the UK, and was in storage at Barrow Hill. It was later moved to the Barry Island Railway, South Wales for restoration.

No. 20301, 23 May 2009

No. 20301 (D8047) is seen undergoing repairs at Eastleigh Works. This loco was rebuilt from No. 20047 in 1995 and had been brought from RFS following Channel Tunnel construction traffic. No. 20301 carries the name *Max Joule 1958–1999*, and is unusual in that it is carried on the cab front.

No. 20301, 18 June 2013

No. 20301 (D8047) is seen at Bescot along with No. 20308 on a nuclear train. The Class 20s had failed the previous night and were waiting rescue. No. 20301 was rebuilt from No. 20047 in 1995 and carries the name *Max Joule 1958–1999*.

No. 20302, 1 June 1996

No. 20302 (D8084) is pictured at Crewe station coupled to classmates No. 20301 and No. 20303, waiting to depart with Pathfinder's Cumbrian Coaster Railtour. This had originated behind No. 37604 and No. 37611, these locos being changed for the Class 20s at Crewe. No. 20302 was rebuilt from No. 20084, which had been brought from RFS following Channel Tunnel construction workings.

No. 20303, 1 June 1996

No. 20303 (D8127) is seen at Crewe coupled with No. 20301 and No. 20302 backing onto Pathfinder's Cumbrian Coaster Railtour. No. 20303 was rebuilt from No. 20127, which had been brought from RFS.

No. 20305, 23 August 1986

No. 20305 (D8172) is seen stabled at Bescot. This was only a short-term renumbering for Peak Forest workings, and it was soon renumbered back to 20172.

No. 20305 and No. 20314, 23 May 2009

No. 20305 (D8095) and No. 20314 (D8117) are seen side by side at Eastleigh Works, both carrying DRS livery. No. 20305 was rebuilt from No. 20095, whereas No. 20314 was rebuilt from No. 20117. Both are still in service today, although No. 20314 now works for Harry Needle and is painted orange.

No. 20305, 17 March 2015

No. 20305 (D8095) passes Longbridge in the Birmingham suburbs, along with No. 20308, on a nuclear flask working heading for Crewe. No. 20305 has received the name *Gresty Bridge* after the DRS depot in Crewe. No. 20305 was renumbered from 20095, and was acquired from RFS following Channel Tunnel construction workings.

No. 20306, 30 April 2008

No. 20306 (D8131) passes Washwood Heath carrying DRS livery while on a nuclear working. No. 20306 was renumbered from 20131, which DRS had brought from British Rail Telecommunications, and was rebuilt in 1998. No. 20306 would be sold to C. F. Booth, Rotherham for scrap, being broken up in 2013.

No. 20307, 23 August 1986

No. 20307 (D8194) rests at Bescot during the summer of 1986. Eight Class 20s were renumbered in 1986 for working in the Peak Forest, but they were soon all renumbered back to their original identities after roughly six months. No. 20307 was renumbered from 20194.

No. 20307, 19 April 2008

No. 20307 (D8050) is seen at Westbury while on a Pathfinder Railtour with No. 20310 to Weymouth. No. 20307 was rebuilt from No. 20128 in 1998, who DRS had brought from British Rail Telecommunications. No. 20307 would eventually be sold to C. F. Booth, Rotherham for scrap, being broken up in 2013 along with Nos 20306, 20310, 20313 and 20315.

No. 20309, 7 February 2009

No. 20309 (D8075) is seen waiting to depart from Carlisle station with a Spitfire Railtours working back to Birmingham, along with No. 37423. No. 20309 had been rebuilt from No. 20075 in 1998. This loco is still in use with DRS today.

No. 20310, 19 April 2008

No. 20310 (D8190) stands on the buffer stops at Weymouth along with classmate No. 20307, having worked a Spitfire Railtour to the town. No. 20310 was rebuilt from No. 20190 in 1998 and carried the name *Gresty Bridge*, which was removed and applied to No. 20305. No. 20310 was one of five Class 20/3s to be sold to C. F. Booth, Rotherham for scrap in 2013.

No. 20311, 15 July 2007

No. 20311 (D8102) is seen at Barrow Hill carrying DRS livery. This was rebuilt from No. 20102 in 1998, and has since been sold by DRS to Harry Needle. It is now painted orange and is used on London Underground tube stock delivery trains.

No. 20314, 31 December 2012

No. 20314 (D8117) is seen at Derby in Harry Needle orange livery. This is one of three Class 20s to carry this livery, No. 20121 and No. 20311 being the others, and the two 20/3s are used on London Underground tube stock delivery trains. No. 20314 was rebuilt from No. 20117 in 1998 for DRS.

No. 20901, 5 October 1995

No. 20901 (D8101) pauses at Clapham Junction with a Hunslet Barclay weed-killing train. Six Class 20s were taken over by Hunslet for these duties and were renumbered 20901 to 20906. No. 20901 was renumbered from 20101 and was sold to Harry Needle, who still owns the loco today, although it is now painted in GBRf livery. No. 20901 was used in Kosovo for a while in 1999 along with No. 20902 and No. 20903.

No. 20901, 15 July 2007

No. 20901 (D8101) is seen at Barrow Hill carrying Railfreight livery. This loco never carried this livery in service; only No. 20088 wore the triple grey livery. No. 20901 was renumbered from 20101 for Hunslet Barclay weed-killing workings in the 1990s, was eventually sold to Harry Needle, and is still in use today on London Underground tube stock delivery workings.

No. 20901, 15 November 2008

No. 20901 (D8101) is seen at Barrow Hill carrying Railfreight livery. The addition of miniature snowploughs really sets the loco off. It is coupled with No. 20905, which is also carrying Railfreight livery, and both these locos are in main line use today carrying GBRf livery.

No. 20901 and No. 20905, 31 December 2012

No. 20901 (D8101) and No. 20905 (D8325) are seen at Derby wearing GBRf livery. They are used on London Underground tube stock delivery trains and the tank wagons are barrier vehicles for the tube stock. The other Class 20s are No. 20314 and No. 20096, and all four locos form part of a large pool of available 20s to work these trains.

No. 20901, 24 April 2013

No. 20901 (D8101) passes Water Orton along with No. 20905 while on a London Underground delivery move. Both locos carry GBRf livery and look very smart. No. 20901 was renumbered from 20101 when it was in use with Hunslet Barclay on weed-killing trains.

No. 20902, 7 June 2009

No. 20902 (D8060) is seen in long-term storage at MOD Long Marston. The loco carries the unbranded DRS livery of its last owner. No. 20902 was renumbered from 20060 when it was sold to Hunslet Barclay for weed-killing trains. When these duties were complete, it was sold to DRS along with the five other Class 20/9s. No. 20902 also visited Kosovo along with Nos 20901 and 20903 in 1999, returning in 2000. It was later sold to Harry Needle and spent time stored at Barrow Hill and Long Marston before being sent to European Metal Recycling, Kingsbury for scrap in September 2011.

No. 20903, 7 June 2009

No. 20903 (D8083) is seen in long-term storage at MOD Long Marston carrying unbranded DRS livery. This was renumbered from No. 20083 by Hunslet Barclay for use on weed-killing trains and would eventually be sold to DRS for further use. This loco visited Kosovo along with Nos 20901 and 20902 in 1999. It was later sold to Harry Needle, and is currently stored at Nemesis Rail, Burton-on-Trent.

No. 20904, 15 November 2008

No. 20904 (D8041) is seen carrying unbranded DRS livery while in long-term storage at Barrow Hill. This was renumbered from 20041 by Hunslet Barclay for weed-killing trains, and was eventually sold to DRS. No. 20904 was later sold to Harry Needle and spent time stored at Barrow Hill before being moved to Nemesis Rail, Burton-on-Trent.

No. 20905 and No. 20901, 31 December 2012

No. 20905 (D8325) and No. 20901 (D8101) are seen at Derby carrying GBRf livery. Both these locos are owned by Harry Needle and are used on London Underground tube stock delivery trains from Derby. No. 20905 was renumbered from 20225 by Hunslet Barclay when it was used on weed-killing trains before being sold to DRS. All six of the Class 20/9s were sold to Harry Needle for further use.

No. 20906, 17 November 2012

No. 20906 (D8319) is seen at Barrow Hill carrying all-over white livery. This loco is owned by Harry Needle and is on hire to LaFarge Cement, in the Hope Valley, and is numbered 3. No. 20906 was originally renumbered from 20219 for use by Hunslet Barclay, but didn't see much use as it was classed as a spare loco. Along with the other five Class 20/9s, it was sold to DRS for further use. This spent many years stored at LNWR Crewe, and also at Barrow Hill, before being hired to LaFarge.

D8001, 7 April 2002

D8001 (No. 20001) is seen at the Midland Railway Centre carrying BR green livery. This view was taken just before it went on hire to CTRL for use on construction trains. It was returned to the Midland Railway Centre and painted BR blue.

D8020, 5 September 1998

D8020 (No. 20020) is seen at the Bo'ness & Kinneil Railway carrying BR green livery. This loco has been preserved for nearly twenty-five years, having been purchased in 1991.

D8057, 21 November 2015

D8057 (No. 20057) is seen inside the repair shed at the Midland Railway Centre, Butterley undergoing restoration. This was finally withdrawn from BR service in 1994 and was stored for many years at Springs Branch depot, Wigan and also at Bescot until it was sold to Harry Needle. It was eventually sold for preservation and has ended up at Butterley.

D8058, 11 October 1968

D8058 (No. 20058) passes through Sheffield Midland with a freight working. This was to lead an uneventful life before being withdrawn in 1992 and sold to M. C. Metals for scrap, which was completed in August 1994.

D8098, 2 August 2007

D8098 (No. 20098) is seen preserved at the Great Central Railway, Loughborough. The loco carries BR green livery with small yellow warning panels, and carries a bodyside ladder, which was removed from most locos early on in their careers. D8098 was preserved in 1992, having finished its BR career at Thornaby.

D8098, 21 November 2015

D8098 (No. 20098) stands resplendent in full BR green livery at the Great Central Railway, Ruddington.

D8110, 11 October 1998

D8110 (No. 20110) is seen at the South Devon Railway carrying BR green livery with small yellow warning panels. The loco has been fitted with miniature snowploughs, which really sets the loco off nicely. D8110 spent all but the last four years of its BR career in Scotland, finally moving south in 1986. This is another Class 20 to have spent nearly twenty-five years in preservation, nearly as long as it was in BR service.

D8132, 16 October 2005

D8132 (No. 20132) stands outside Barrow Hill having been repainted into BR green livery, but with small yellow warning panels, and the headcode roller blinds had been refitted. This loco lead an uneventful life, but was hired out for Channel Tunnel construction trains and sent to France. Upon return to the UK, it was sold to Harry Needle for spares, and was then sold to DRS for the same purpose before being sold back to Harry Needle. It was later restored to main line use and can be seen on London Underground tube stock delivery trains, repainted into red stripe Railfreight livery. It now carries the name *Barrow Hill Depot*.

D8154, 11 September 2005

D8154 (No. 20154) is seen on display at the Adtranz Crewe Works open day. The loco carries BR green livery, but with small yellow warning panels, and also carries black miniature snowploughs. This was originally saved for preservation at the Churnet Valley Railway, but has since moved to the Great Central Railway, Ruddington.

D8188, 17 May 1998

D8188 (No. 20188) is seen on display at Bournemouth open day in 1998. At this time the loco was owned by Waterman Railways and carries their black livery and the name *River Yeo*. It was later sold for preservation, and can currently be found at the Severn Valley Railway, owned by the 72C Traction Group.

D8198, 27 May 1967

D8198 (No. 20198) passes through Derby station light engine, carrying BR blue livery, when only one month old. This loco spent many years working north of the border before being withdrawn in 1991. It was sold to M. C. Metals, Springburn for scrap, which was completed in October 1993.